The Canal du Midi

Translated from the French "Le Canal du Midi" by Julian Guest

First published in the United States in 1983
by Thames and Hudson Inc., 500 Fifth Avenue,
New York, New York 10110

© 1981 Editions Technal International, Toulouse

Library of Congress Catalog Card Number 82-50804

ISBN 0-500-24115-5

Printed and bound in Italy

The Canal du Midi

Text by Odile de Roquette-Buisson

Photographs by Christian Sarramon

Captions by Isabel Lefebvre

with 164 photographs, 141 in colour

THAMES AND HUDSON

The canal-boat harbour in the roadsteads of Bordeaux. (*Preceding page.*)

(*Left*) Constructional details of the canal architecture.

(*Right; top to bottom, left to right*) Steps at the Fonserannes locks; shallow arched bridge (19th century); "Pont Riquet", near Homps (17th century); the Repudre aqueduct; single lock, typical of many on the canal; the Puicheric aqueduct; bridge and lock "de l'Océan"; Pont de Rieux; three-arched "Pont Riquet", near Agde; installation at Libron; weir on the Laudot; a single lock seen from the air.

The Cesse lock on the Sallèles-d'Aude branch. (*Overleaf*) To the left of the bridge is a conduit conveying a permanent supply of water to the surrounding countryside. Such diversionary channels were built between 1900 and 1910 in order to wash away the phylloxera infesting the vineyards.

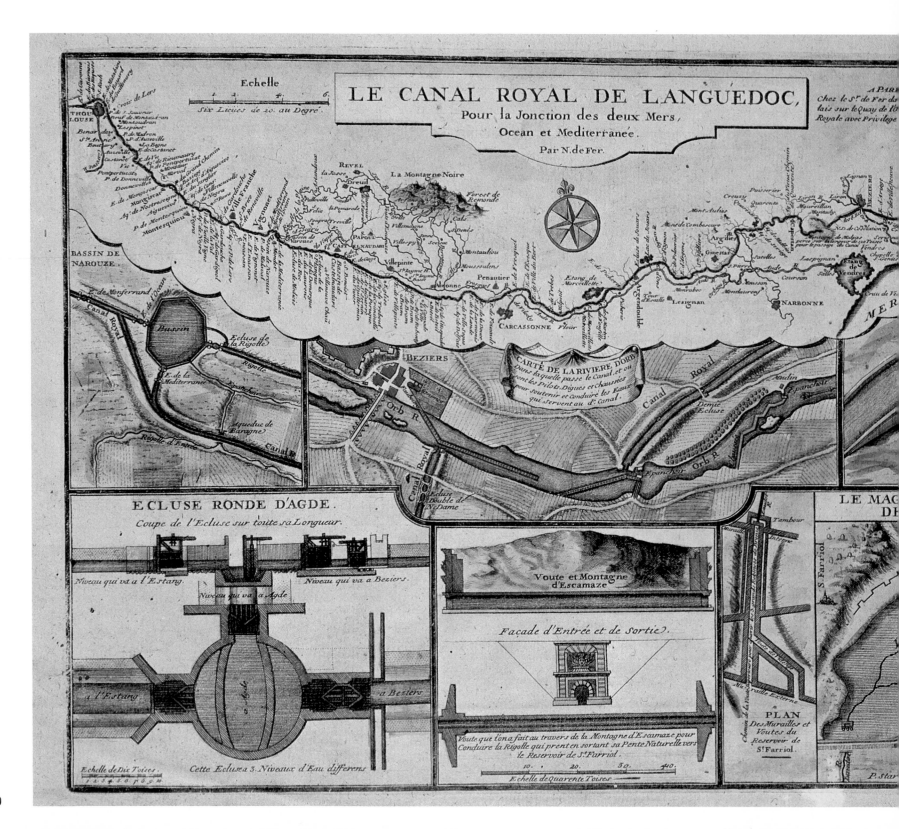

LE CANAL ROYAL DE LANGUEDOC,
Pour la Jonction des deux Mers,
Ocean et Mediterranée.
Par N. de Fer.

Echelle

Six Lieües de 20. au Degré.

BASSIN DE NAROUZE.

ECLUSE RONDE D'AGDE.

Coupe de l'Ecluse sur toute sa Longueur.

Niveau qui va a l'Estang.

Niveau qui va a Agde.

Niveau qui va a Beziers.

a l'Estang

a Beziers

Echelle de Dix Toises.

Cette Ecluse a 3. Niveaux d'Eau differens

Voute et Montagne d'Escamaze

Façade d'Entrée et de Sortie.

Voute que l'on a fait au travers de la Montagne d'Escamaze pour
Conduire la Rigolle qui prend en sortant sa Pente Naturelle vers
le Reservoir de St Farriol.

Echelle de Quarante Toises

PLAN
Des Murailles et
Voutes du
Reservoir de
St Farriol.

CARTE DE LA RIVIERE D'ORB
Dans laquelle passe le Canal, et où
vont les Pilots Digues et chaussées
pour soutenir et conduire les Eaux
qui servent au d't Canal.

BEZIERS

This map is a
remarkable document
dating from the year
1726. The upper part
shows the canal from
Toulouse to the
Mediterranean, all the
towns and villages
along the route being
marked. In the insets
below are the plans of
various works, some of
which have disappeared
or been altered since
Riquet's day: the dock
at Naurouze; the
crossing of the Orb at
Béziers before the
construction of the
aqueduct; the Repudre
aqueduct; the circular
lock at Agde; the tunnel
through the Cammazes;
the dam at
Saint-Ferréol. (Archives
of the Canal du Midi.)

IT IS THE SIESTA HOUR. The reflections of the hot afternoon sun
glitter and sparkle on the surface of the water, filtered by the thick
foliage of the plane trees overhead. A barge glides slowly down
the canal towards the next lock after Ayguesvives; it is called
"l'écluse du sanglier", the lock of the Wild Boar.

Marius, the boatman, is telling us how he inherited his barge
from his father. She is called "la belle Isaure", a name he likes to
roll around his tongue although there is a twinkle in his eye as he
tells us that he has no idea why his father named her thus.

Today is the 15th of May and, although Marius does not realize
it, a great anniversary for the Canal. To tell the truth Marius could
tell you very little about the history of his canal except perhaps
that someone called Riquet built it. Marius loves his boat and his
way of life; he senses rather than sees the beauty of his leisurely
water route which leads from one sea to another.

Marius and his boat are timeless. They could equally well have
formed part of the string of barges from the Garonne which
followed the glittering procession of the great men of France,
who on 15 May 1681 came from far and wide to inaugurate "le
Canal Royal du Languedoc" and made their way with much
ceremony to the great fair at Beaucaire. The idea of a canal to
link "the two seas", a cherished dream ever since the reign of
Henri IV, had at last been fulfilled by a local man, Pierre Paul
Riquet, after twenty arduous years of constant endeavour.

Our present-day sailor is not a man of great education. He is
happy in his work and does not bother his head overmuch about
his forbears. Nevertheless, you have only to place him on the
towpath in seventeenth-century attire and with a horse, and you
will immediately see in the mind's eye his direct ancestor, who
would have spoken to you with the same simple directness and
the same thick accent.

11

Pierre Paul Riquet, hero of a forgotten province

THE OLD PROVINCE of Languedoc was the largest area in France. Blessed with a soft climate almost untouched by winter frosts, it produced, then as now, an abundance of corn, fruit and oil-bearing crops and at one time the highly prized woad which was used in the dyeing of cloth. In the even milder Mediterranean hinterland the crops are olives, saffron, mulberries and wine. Many were the successful businesses built up on the foundation of this agricultural economy.

Unfortunately, in the 17th century roads and bridges were in an appalling state and infested with brigands. Exporting all this produce to the outside world thus became an extremely risky undertaking. The alternative sea route round the coast of Spain was if anything more dangerous still, on account of the ravages of pirates and storms. The cost of hiring vessels capable of carrying produce as far as the Straits of Gibraltar for sale or barter there was not really an economic proposition.

Riquet, a man of humble origins, felt the same bitterness towards and sense of alienation from the Kingdom of France as any peasant from the Midi. Oral tradition kept alive the memory of the terrible sufferings of the Languedoc people at the time of the Albigensian Crusade of the 12th century and of the recent religious wars. The northern lords with their strange unmelodious language and their hated taxes, vaunting themselves over the conquered southerners, were only accepted outwardly and then only by force of arms. Often they had even taken over the fine châteaux of the ousted southern seigneurs; of such families as the Raymonds, Counts of Toulouse, of the Trencavels and the Cabarets, whose proud deeds lingered on in fireside tales of brave exploits. In spite of the goodwill of King Henri IV, himself a southerner from the Béarn, the men from the banks of the Garonne felt in their hearts a deep sense of injustice which could only be healed by a new epic achievement in their province — the creation of the Canal Royal du Languedoc, that would at last enable all those who could not feel French to be proud of being so. Through the intervention of Riquet, they would recognize their sense of belonging and thus accept their King.

In 1666 the King proclaimed in the preamble to the edict authorizing the building of the canal: "This will be a great work of peace, worthy of our diligence and care, capable of perpetuating throughout the centuries to come the memory and greatness of its author and the abundance and felicity of our reign."

The King's spark of genius and the meeting of two minds who shared a passion for building provided the impetus which made of the sparkling narrow ribbon, winding its way through the fields and hills of the South, the gentlest of links joining the great province to the rest of the Kingdom.

Who would have recognized in the solitary man strolling along the footpaths at Arfons, the mighty *"entrepreneur-fermier des gabelles et fournitures aux armées du Roi en Cerdagne et Roussillon"*? For he was no other than Riquet who in his fifties had fallen in love; in love with the deep cool forest with its softly carpeted bracken thickets under the tall pines and its tinkling crystal waterfalls. Ramondens, the château which he had acquired at the foot of the Montagne Noire, rejuvenated him after the years of unremitting toil in his lucrative but thankless task as tax collector. During the first part of his life he had taken advantage of his privileged position to achieve material and social success. His considerable fortune, thus acquired, now gave him the leisure to live with his family, spending his days between his property of Bonrepos, near Toulouse, and his beloved Montagne Noire.

Nothing escaped Riquet's eagle eye, whether during his journeyings on official business in the territories of the Cerdagne and Roussillon, newly acquired by France, or among the vineyards and dried-up watercourses of his birthplace, the Biterrois, or

Pierre Paul Riquet (1609-1680), the instigator of the canal. Among other signs of gratitude, Louis XIV ennobled him with the title Baron de Bonrepos.

indeed on the roads between Carcassonne and Toulouse, the capital of the province, in the course of his visits to Ramondens and Bonrepos. We know little of him as a person, but it is more than probable that he will frequently have had to listen with growing displeasure to the complaints of his underlings about the treatment they received at the hands of the local peasantry who were almost in a state of open revolt, so harassed did they feel by heavy taxation.

All in all, the life of the people of Languedoc at the beginning of Louis XIV's reign was becoming more and more difficult. What with the malarial swamps spreading along the Mediterranean coast, the harsh winds whistling through the sails of the windmills that ground the woad — legendary wealth of the Lauragais, which was rapidly being supplanted by the indigo trade — nothing seemed to be going right. The heavy clay made the hill roads dusty and rutted in summer and a slushy mire in winter. It was almost impossible to move merchandise in quantity under such conditions. As for the valley roads, dominated by the frowning Corbières, there was always the danger here of attack by Spanish brigands.

Riquet, in the peace of his estate at Ramondens, reflected on the miseries of his province. Perhaps he owed his passion for building to his Florentine ancestry and, like Leonardo da Vinci before him, he dreamed of opening up a waterway by extending the river Garonne as far as the Mediterranean. He was not the first to entertain the idea. The Romans had already considered it and, more than a century before, Nicolas Bachelier and later Pierre Roneau, in the reign of Henri IV, had made much progress in planning a canal that was to link the two seas.

All these projects were abandoned, however, for the following reasons. For four or five months of the year the streams flowing from the Montagne Noire yield insufficient water to fill the canal

13

The château of Bonrepos, Riquet acquired the
lands of Bonrepos, a few kilometres north of
Toulouse, and commissioned the contractor
Roux to build him a château in the style of the
aristocratic Toulousain houses of the
17th century. Here on his estate he
experimented with all the hydraulic systems of
the canal in miniature.

from a key point known as the Stones of Naurouze. In order to solve this problem it was suggested that channels joining the rivers Garonne and Aude should be built, but it was soon decided that this would be too expensive. Riquet himself remembered hearing his father Guillaume saying, at the time when the Estates of Languedoc turned down a proposition by a man named Bernard Arribat, that it was a madcap scheme and that "it would be detrimental to both the private and public weal".

But the idea continued to haunt Riquet himself. There *was* water; plenty of it. He could hear it chattering and see it glistening on the rocks in the summer sunlight. Near his Château of Ramondens small streams and rivers abounded — streams such as the torrentuous Alzau, the Bernassonne, the Lampy and its small tributary the Lampillon, the Rieutort and not least the Sor. This far-seeing man, inextricably bound up by virtue of his office with the economic life of his province, knew that a canal, contrary to his father's stated opinion, would be "valuable to both the private and public weal". He had only to observe commercial traffic in other parts of France; this type of thoroughfare suited the age and the men of the age, for the most part good sailors, who found it perfectly logical to extend the highways of the sea by means of inland waterways.

One evening while turning over in his mind the problem of diverting the waters so that instead of flowing towards the Mediterranean they should flow in the opposite direction, towards the Atlantic, Riquet hit upon the solution. It was obvious: build a sort of trench (*rigole*) round the side of the mountain and the waters thus collected could be channelled to Le Conquet where they would merge with the river Sor. All that needed to be done then would be to continue the *rigole* across the plain to the canal.

Riquet had his horse saddled first thing next morning and hurried to Revel to see Pierre Campmas, the well-digger's son. He was certain that Pierre would endorse his solution to the problem and in any case he would need Pierre's advice concerning the techniques to be used.

Since he was a provincial and far removed from the court, Riquet had above all to find some powerful support to help advance his project. Luckily, he was friendly with Monseigneur d'Anglure de Bourlemont, Archbishop of Toulouse. Once the latter had been convinced that Riquet's solution was the correct one, he explained to him the best method of getting a hearing. As a result, on 15 November 1662, Riquet sent his famous memorandum to Colbert, the King's chief minister. The great undertaking had begun !

The Great Venture: linking the two seas

IT IS EASY TO IMAGINE Riquet's anxiety while he waited for permission to start his great work, after having got thus far in 1662. The Pyrenees showed white, slicing across the horizon, as Riquet pondered. The treaty of the Pyrenees, which had made the young King bow to reasons of state and marry Maria Theresa, Infanta of Spain, and renounce his true love, the charming Marie Mancini, niece of Cardinal Mazarin, had raised France to the highest pinnacle of power it had ever known. After a century of struggle it had wrested supremacy from Austria and now dominated the Christian world. An heir (the Dauphin) was born of this marriage and the population greeted his arrival with such a spate of rejoicing that the bells and cannon could hardly be heard above the cheering crowds; just like twenty years before when Louis XIV was born.

Riquet remembered that occasion well. The union of Louis XIII and Anne of Austria had seemed to be a complete frost. For twenty-three years the people had been waiting to hear the news of the birth of an heir, but everybody knew that as long as the Queen lived in Paris the King would refuse to go there. While she pined in the Louvre, he went stag-hunting in the deep forests which surrounded the then small hunting lodge of Versailles. Only a terrible storm would have persuaded Louis to spend the night in his hated wife's bed. Were the prayers of Mlle de Lafayette, the King's favourite, now a young novice in a convent in the rue Saint-Antoine, who had just received a royal visit and had had a long interview with him, perhaps fulfilled in this manner? This sweet young woman still had a strong hold over him even though she had fled from his unworthy importunities and taken refuge in the convent of the Order of the Visitation. On this evening of 5 December, maybe he stayed so long talking to her that to have returned to his usual refuge, the home of the Prince of Condé, would have been dangerous at that time of night in that weather. Whatever the reason, he definitely stayed the night at the Louvre and nine months later to the very day, Anne of Austria was delivered of the child of that storm — the future Louis XIV.

Indeed a remarkable entry into the world for the Dauphin, an event which nobody had any longer believed would happen. He would be King at five years old. He would be proclaimed of age at thirteen and immediately announce his intention of taking over the reins of government himself. In thanking his mother the Queen for surrendering to him the powers she had held as Regent during his minority, he desired her to "have the kindness to continue to give him good counsel". At sixteen he was crowned at Reims. In future it was to God alone that he would be accountable "for that power which no Frenchman could legally disobey".

Louis XIV was a handsome man, despite the ineffaceable marks of smallpox, which had nearly carried him off in childhood. "His hard tight-lipped mouth, although sensual, warns the onlooker not to rely on his soft caressing look," said a contemporary. Such was the King on whom all enterprises including peace and war depended, according to his pleasure. Pierre Paul Riquet, however, was convinced that, as soon as he had persuaded the King's principal Minister, Colbert, of the good sense of his plans, he had as good as obtained the flowing "Louis" of the King's signature on the edict authorizing the project so dear to his heart. It gave him the feeling that his whole life had been nothing but an introduction to this incredible venture which he was certain he could carry off. Like his King, he was one of those men to whom the future belongs and who knows how to stake his reputation and fortune on a single throw.

It took more than twenty years for Riquet to make this cherished French dream come true. Riquet, with technical assistance from Pierre Campmas, proposed making a trial *rigole* complete with stopcocks and even a real mountain with pipes tunneled through it. At Bonrepos, therefore, they built a working model. It turned out to be so successful that they decided to present the scheme to the Contrôleur-Général des Finances, Colbert.

Colbert presenting the plan of the "Canal du Languedoc" to Louis XIV. In this document, the Canal du Midi takes its place among the great projects advanced by the King's leading minister. Soon convinced of the strategic and economic benefits to be reaped by a canal linking the Atlantic with the Mediterranean, he whole-heartedly supported Riquet's scheme. (Musée Paul Dupuy, Toulouse.)

COLBERT PRÉSENTE A LOUIS XIV

LE PLAN DU CANAL DE LANGUEDOC.

Jean-Baptiste Colbert, Marquis de Seigneley, &c. né à Paris en 1619. S'étant distingué par ses talens et sa fermeté, le Cardinal Mazarin le connut, et lui inspira à tellement la confiance de Louis XIV, qu'à la mort du Ministre, cette confiance se trouva toute établie. Mazarin le recommanda au Roi comme un homme d'une application infatigable, d'une fidélité à toute épreuve, et d'une capacité supérieure dans les affaires. Devenu Contrôleur-Général après la disgrace de Fouquet, il rétablit bientôt l'ordre, améliora l'état des finances, protégea le Commerce, les Lettres et les Arts. Le Roi le nomma, en 1664, Surintendant de ses Bâtimens: Colbert fit revivre les Arts qui ont rapport aux Bâtimens; la France vit des chef-d'œuvres de Peinture, de Sculpture, d'Architecture, tels que la façade du Louvre, la Galerie de la colonnade, les écuries de Versailles, l'observatoire de Paris: il se forma, sous ses auspices, de nouvelles Sociétés de Gens de Lettres et d'Artistes: l'Académie des Inscriptions prit naissance dans sa maison en 1663; celle des Sciences fut érigée trois ans après, et celle d'Architecture en 1671. Les Savans dans tous les genres se ressentirent de la protection que le nouveau Mécène accordoit à tous les Arts. Non content d'avoir rétabli l'ordre dans les finances, Colbert porta ses vues sur la Police, le Commerce et la Marine: on vit paroître des règlements utiles; le Commerce fut généralement cultivé; le Canal de Languedoc, dont le projet avoit été formé par Paul Riquet pour la communication des deux mers, fut construit; les Vaisseaux se multiplièrent; les arsenaux furent pourvus de tout ce qui étoit nécessaire à l'armement et à l'équipement de plusieurs flottes. Chaque année de son ministère fut marquée par l'établissement de quelque manufacture; on fabriqua en France un grand nombre d'objets que les Etrangers nous vendoient auparavant, et ce furent autant de conquêtes sur l'industrie des nations voisines. Le but de ce grand homme étoit d'enrichir le royaume et d'accroître sa population. Dès son entrée dans le Ministère, il avoit fait remettre trois millions de tailles, et tout ce qui étoit dû d'impôts depuis 1647 jusqu'en 1656. Telles étoient ses occupations continuelles, lorsqu'il mourut le 6 Septembre 1683, à l'âge de 64 ans six jours, consumé (dit un Historien) des chagrins que lui donnoit Louvois en le forçant à ruiner; par des vexations, le peuple qu'il avoit enrichi par le commerce. Cette fâcheuse position fut sans doute la source des clameurs qui s'élevèrent contre lui. Ses enfans lui firent élever un superbe mausolée dans l'église de St. Eustache.

A Paris chez Blin, Imprimeur en Taille-Douce, Place Maubert, N.º 17, vis-à-vis la rue des 3 Portes. A.P.D.R.

They could not have found a better intermediary to convince the King of the merits of their plan. Louis had the greatest confidence in the judgment of his austere minister, who, passionately interested in anything which would encourage the expansion of commerce, was doing everything in his power to increase the prosperity of France. Riquet was clever enough to put his case in such a way as to capture Colbert's imagination. "The facility and the assurance of this navigation", he said, "will cause the Straits of Gibraltar to cease to be a necessary passage and the revenues of the King of Spain will thus be diminished, while those of our King will be increased by a similar amount."

The wheels of officialdom in those days, as in our own, turned slowly. The decree from the Royal Council was only issued a year later, on 18 January 1663. A commission was appointed, which, joined by another set up by the Estates of Languedoc, would examine the route to be followed. But this commission began to sit only in 1664. Three years lost! Riquet, however, continued to push ahead and plan. Now that he had the "ear" of the powers-that-be, he would hold on to this advantage.

Helped by Pierre and the contractor Roux, Riquet charted the route to be taken by the mountain trench from Durfort to the Stones of Naurouze. Then, with M. de Bourgneuf, who had designed the Canal de Briare, he also marked out the route of the canal proper. At last the famous commission arrived, and sat from 8 November 1664 until 17 January 1665. Among them was one Andréossi, of whom we shall hear more later. After looking at the plans and studying the estimates, the commissioners gave a favourable verdict; "It would appear that the canal is a feasible proposition and that it would be possible to lead it to the lagoon at Thau from where a communicating channel could be opened up with the port of Sète."

The report of the experts was approved, but the commissioners, ever prudent, "desire and believe that a *rigole* two feet wide should be constructed as a trial, to allow a small stream from the

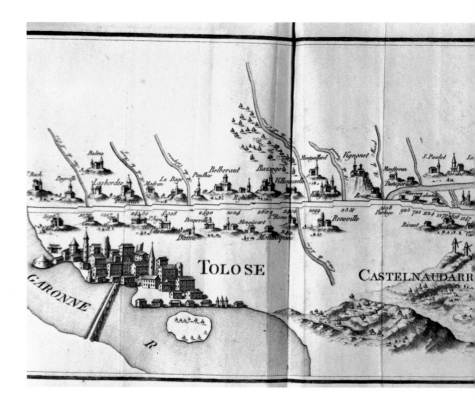

river Sor to flow to the watershed between Toulouse and Carcassonne." Riquet wrote to Colbert suggesting that he build the trial trench at his own expense. Colbert agreed to this, although Riquet had to make many journeys to Paris in order to obtain the necessary letters-patent authorizing the work. The task was begun under the inspection of Messieurs de Tubeuf and Bezons. It was finished towards the beginning of August 1665 and was a complete success. Colbert himself wrote to Riquet; "I am extremely pleased to see your hopes in the grand design of the joining of the two seas crowned with success. As it is you who have resurrected this idea in our age, you need have no doubt that, as

This, one of the first maps of the canal, was requisitioned by Riquet in 1664 at the very beginning of the work. The course of the canal and the means of supplying it with water are only roughly indicated. (Musée Paul Dupuy, Toulouse.)

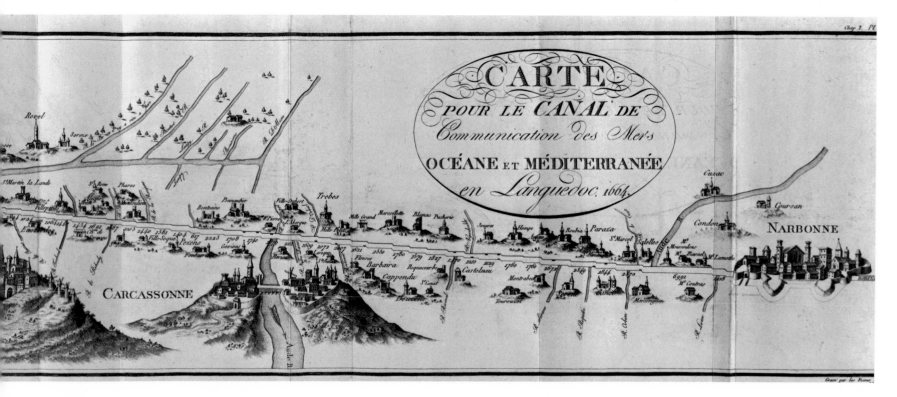

well as the glory which you will reap, the King himself will be extremely grateful."

Nevertheless, as such a huge undertaking could not be carried out without incurring considerable expense, the then governor-general of Languedoc, the Prince of Conti, proposed to the Estates General that the province should contribute to the costs. It has often been said that nobody is a prophet in his own country. Riquet met with great distrust and the Estates declared themselves "unable at present, or in the future, to participate in the cost of building the canal". Nor could Riquet count on the royal coffers. Since his earliest childhood Henri IV's grandson had

known that his reign would be measured by the manner in which he waged war. Eight years of peace had become unbearable for the French nobility and Louvois, the young minister for war, was burning to show his organizational talents. Also Turenne was preparing his plans for the Flanders campaign and the royal treasury was empty. So once again Riquet suggested that he carry out the work at his own expense, if the King would agree. At this point the latter could hardly do less than appoint the Chevalier de Clerville, the Commissioner-General of Fortifications, to prepare an estimate "of the works to be carried out between Toulouse and Trèbes". In the event de Clerville adopted the whole of the plan

19

proposed by Riquet, changing only the dimensions proposed by the experts. At last the King's proclamation was read out in October 1666.

Many interesting points are raised in this edict : the compensation due to the seigneurs whose fiefs the canal crossed and their right to construct châteaux and mills on the banks of the said canal ; the exemptions from tolls and the fishing and hunting rights. There is also mention of the sole rights to the construction of barges and the appointing of officers of justice and "twelve guards in the King's livery".

The undertaking brought great prosperity in its train. Notices concerning the canal were published in the two principal towns of the province. In order to get things moving as quickly as possible, Riquet spread operations over several different sites, each having its own headman and 250 workpeople, with a leader in overall charge. In order to attract the considerable workforce needed for such a large undertaking (some 11,000-12,000), anyone between the ages of 20 and 50 (both men and women) was taken on, and the pay was good. Those on piecework received for each stint a small zinc token struck with the half fleur-de-lys of Riquet's coat of arms. The workforce being predominantly of rural origin, there were often interruptions at harvest time and during grape-gathering. We can readily picture the heavy loads of stone brought to the scene by peasant carts during the construction of roads and bridges and imagine the womenfolk carrying their baskets of rubble in Indian file like a column of ants. This canal was to be the handiwork of an entire province and each completed stage was celebrated with great jubilation.

In April 1667 the first stone of the dam at Saint-Ferréol was laid in the valley of Vaudreuille. A great ceremony was held in the presence of the Archbishop of Toulouse and the Bishop of the neighbouring abbey of Saint-Papoul. Another was held at the laying of the foundation stones of the lock connecting the canal with the Garonne. One stone was laid by the Parlement and the other by the Archbishop ; this carried the inscription : "In the reign of our august sovereign, Louis XIV and under the auspices of his enlightened minister Jean-Baptiste Colbert, this stone, blessed by the worthy Archbishop Charles d'Anglure de Bourlemont and destined to support the enormous mass of the canal which will join the two seas, was laid by Gaspard Lafaille and Pierre Maynial at the request of Pierre de Riquet."

It is now time to return to the part played by commissioner Andréossi. This remarkable young engineer had had the opportunity to study canals and the working of locks in Italy. Hence his friend Riquet employed him on the construction of the canal for which he was responsible. Riquet claimed that he himself had only "a small understanding of mathematics and had studied neither Greek nor Latin". This astonishing man knew his own limitations and had the good sense to surround himself with first-rate assistants, of whom Andréossi was one. What a pity, then, that in 1670 the latter published under his own name, with a dedicatory epistle to the King, a "Map of the Canal of Languedoc", the gist of which was wholly disavowed by Riquet. Today we would say that Riquet had been "had" and felt rather bitter about it. Moreover, more than a century later, in year 8 of the Republic, a descendant of Andréossi tried to attribute the success of the canal's construction to his ancestor alone. This was a needless quarrel, which only served to tarnish the reputation of an undoubtedly great engineer. We can safely leave the last word to La Fontaine : "Gold may be shared, but not praise !"

The bitter flavour of the venture, the triumph of tenacity

DURING THE YEARS 1668-70 the execution of "useful projects whose success rests upon a state of peace" was endangered by the War of the Spanish Succession. Riquet, however, badgered Colbert and pestered de Clerville until he succeeded in getting the second part of the estimate agreed. This part concerned the stretch of the canal onwards from Trèbes to the lagoon at Thau. M. de Bezons duly received the order to sign a contract and Riquet remained the nominated contractor. He obtained 1,200,000 livres from the King for the work. He also managed to have himself put in charge of the farms controlling the Cerdagne forges and had the mines reopened so that the great quantity of iron needed for the mechanical fittings of the locks could be produced. At last one part of the canal was finished.

In November 1670, M. de Seignelay, Colbert's son, visited the new reservoir at Saint-Ferréol, the mountain *rigoles* or trenches and the canal itself from Toulouse to Agde. He was astounded by the enormous amount of work that had been carried out. The port of Cette (Sète) attracted his particular attention. It was already a thriving town with a newly built harbour breakwater, a church, a well, warehouses for provisions and powder magazines, as well as living quarters for Riquet's workmen and stabling for 1200 horses. He celebrated Christmas there, together with the members of the Estates of Languedoc. All in fact was progressing smoothly had it not been for the very great difficulty that Riquet had in obtaining the necessary funds to continue the work. In the end he was obliged to borrow at ruinous rates of interest. He complained frequently to Colbert : "Messieurs de Bezons and de Pennautier agree that I am indeed unfortunate in having found the way to change the course of streams, even though success rewarded me ultimately." In another letter he wrote : "I regard this work with the same affection as I do the dearest of my children and it is indeed true that, even with two daughters to set

Towards the beginning of 1672, it took less than six days for the waters of the trench to fill that part of the canal which ran from Toulouse to Naurouze. Four of the largest of the Garonne barges made the journey and returned laden with cargo. The wine-growers of Gaillac, who had previously been unable to sell their wines in the Bordeaux region, set up a regular thrice-weekly service to bring barrels to Languedoc. It would appear that if the canal had been completed as far as the Mediterranean that year, a good business selling corn to Italy could have been inaugurated, as that country was in the grip of a terrible famine.

Towards the end of 1672, Pierre Paul Riquet fell ill. He had embarked on his colossal undertaking at the age of fifty-three and ten years of incessant struggle had exhausted him. With his customary foresight he had anticipated this very possibility and had arranged that his eldest son Jean-Mathias Riquet de Bonrepos should replace him as contractor, in the event of his death. However, his hour had not yet come, and he recovered.

Notwithstanding his return to health, Riquet still had his problems. The inhabitants of Carcassonne, less far-seeing than up in the world, I prefer to keep them with me for a while longer and employ for the continuance of my work the moneys previously set aside for their dowries."

To add to these financial problems, Riquet aroused much enmity and envy, and many calumnies were heaped upon his head. Nevertheless his faith in the ultimate success of the great enterprise never faltered. Moreover, he was granted moments of encouragement, as when the inhabitants of Castelnaudary begged him to allow the canal to pass through their town. Thanks to a subsidy paid by the town itself, Riquet's agent Contigny arranged for the canal to be routed through the place called "pré de l'étang", where one can still see the charming little port that was subsequently built.

21

the "Chauriens" (as the men of Castelnaudary are called), refused to contribute towards the cost of running the canal within their town walls. The Malpas incident also needs to be mentioned. Was it inspired by doubt as to the efficacy of Riquet's ideas, or simply by jealousy? It is difficult to be certain, but Riquet was almost left in the lurch by his colleagues. The dispute grew up over the method to be used in crossing a major obstacle in the way of the canal — the river Ognon. M. de Clerville disagreed with the calculations of Father Mourgues; the former wanted to overcome this obstacle by building an aqueduct linking a natural chain of rocks that formed "stepping stones" across the river, thus keeping the canal level and in the plain. Father Mourgues pointed out that if this idea was adopted they would then come up against the problem of crossing the river Aude, an all but insoluble one in view of the limited techniques of their age. Much time was lost in wasteful argument. At this point Riquet intervened; the canal would take the high route, even if it meant cutting a tunnel through the Ensérune hill, a procedure which his specialists regarded as unsafe, as they suspected that the hill was formed of porous chalk. Unbeknown to them both, however, Riquet ordered Pascal de Nissan to bore a tunnel four feet wide through the hill. This proved such a success that the Malpas tunnel was inaugurated by the light of torches in the presence of Cardinal de Bonzy, Archbishop of Narbonne, and of the King's Commissioners.

In spite of the war with Holland, which lasted six years and employed all the resources of the kingdom, the work went on as before, with Riquet risking his own capital in the venture. Already the port of Sète offered real protection to shipping and the King showed his gratitude by granting certain privileges and rights, including the right "in perpetuity" for M. de Riquet and his descendants "to fish in the port and in the canal when opened as far as the lagoon of Thau".

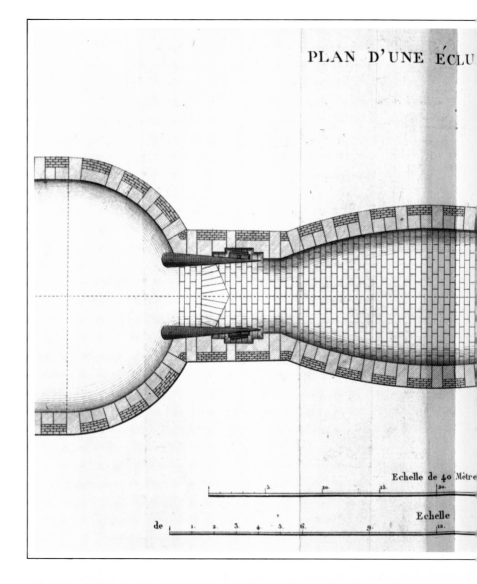

PLAN D'UNE ÉCLU

Echelle de 40 Mètre

Echelle

Plan of a typical lock, together with details of the gates and mechanism. The 102 locks on the canal were all constructed to the same plan. Their "olive-shaped" form was adopted for technical reasons. Firstly, it extended the area of the lock itself; secondly, it reduced the pressure exerted by the surrounding earth which might otherwise have ultimately damaged the walls. Above Riquet's lock gates was a large horizontal bar which the lock-keepers operated by hand. The wooden gates soon rotted, however, and needed constant replacing. At a later stage cast-iron was gradually substituted for wood.

(Musée Paul Dupuy, Toulouse.)

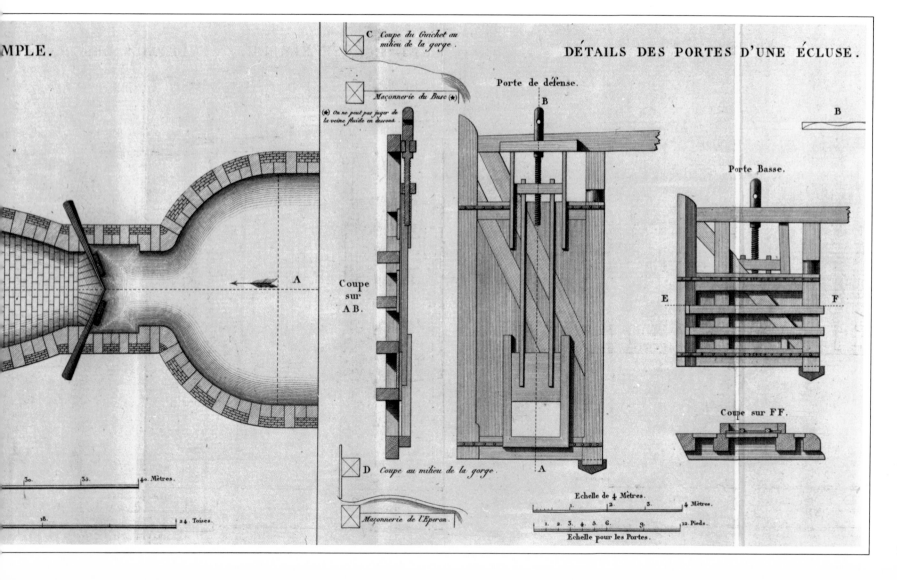

MPLE.

C *Coupe du Guichet au milieu de la gorge.*

Maçonnerie du Busc (★)

(★) *On ne peut pas juger de la veine fluide en dessous.*

Coupe sur AB.

D *Coupe au milieu de la gorge.*

Maçonnerie de l'Éperon.

DÉTAILS DES PORTES D'UNE ÉCLUSE.

Porte de défense.

Porte Basse.

B

E   F

Coupe sur FF.

Echelle de ¼ Mètres.

Echelle pour les Portes.

Sadly, at the very time when the entire length of the canal was about to become navigable, Riquet died on 1 October 1680. The torch was taken up by his son, Jean-Mathias, who begged the King to order a final inspection of the work. Three commissioners were appointed: Messieurs d'Aguesseau, de la Feuille and Mourgues. They left Béziers on 2 May 1681 and followed the route of the canal as far as the junction with the river Garonne, using the towpath. Everything was inspected in detail; embankments, bridges, sluices and towpath. The same minute inspection was made of the two *rigoles* and the reservoir at Saint-Ferréol. M. d'Aguesseau gave the order to fill the canal with water and, once filled, the commissioners boarded a large barge. Soundings were made and a report on the state of the canal issued.

Spring 1681 saw the grand inaugural procession. A religious ceremony was conducted by Cardinal de Bonzy; then the bishops, the King's commissioners and the members of the Estates of Languedoc took their places on spendidly decorated barges. A galley led the procession, and another vessel carried an orchestra; this was followed by kitchen and pantry boats. Finally came the large barges from the Garonne, laden with merchandise from France, England and Holland, all destined for the great fair at Beaucaire. Then the Royal Canal of Languedoc was officially opened, to the acclamation of a multitude of people from both town and countryside, who had come to form a guard of honour to celebrate the birth of this prodigious handiwork which they felt in a very real way to be their own.

A breath of fresh air, laden with the salty breezes of the Mediterranean and the Atlantic, had brought hope to the men of the South. All along the canal, watching the procession go by, they raised a cheer : "Long live our good King Louis XIV, a fair wind to our boats and good luck to our merchandise."

In order to better understand what the inventor of the canal had to contend with, it is useful today to study the numerous documents which inform us about the state of Languedoc in the 16th and 17th centuries. After the bitter struggles of the Cathar or Albigensian crusade, with the resulting death of the young King Pierre II of Aragon, killed before the walls of Muret while aiding his brother-in-law, the Count of Toulouse, still fresh in the memory, southern independence seemed to have been dealt a fatal blow. It was as though the province had been conquered by an enemy. "One may say that the men of Languedoc do not wish to be dissolved in the French melting pot," wrote the chronicler. The southerners continued obstinately to speak their own language and despised from the bottom of their hearts the lands above the Garonne, where the earth is only warmed by a pale sun and their own favourite crops cannot be grown because of the winter frosts. Olives, mulberries, woad and millet were all as essential to the way of life of the South, as were the warm winds whose strength seems to diminish when they reach the Garonne valley, as if they too wished to remain part of Languedoc. The ills the people suffered were, however, exactly the same elsewhere in the Kingdom. There was disease and pestilence, the most devastating plague having struck in the years 1617, 1628, 1629 and 1652, dates that are remembered to this day. Before God's flail the people prayed for deliverance in their ancient tongue :

Prenez, Grand Diou compassioun
de naustres et de nostre villo
Tout s'en vai daja filo à filo
vostre pople tout dessequat,
semble que l'agoun enmasquat
et noun pas cap de famillo
qu'oun age perdut fil ou filo... (Daniel Le Sage)

Plans of the sequence of locks at Fonserannes and of the tunnel at Malpas. By introducing a "stairway" of seven locks at Fonserannes, which climbs the steepest part of the whole canal system, worthwhile economies.

in masonry and material for lock gates could be made. At Malpas, the 198 km mark, a small tunnel 160 m long was dug in the chalky subsoil of the Ensérune hill, thus enabling the canal to bypass the highest point.

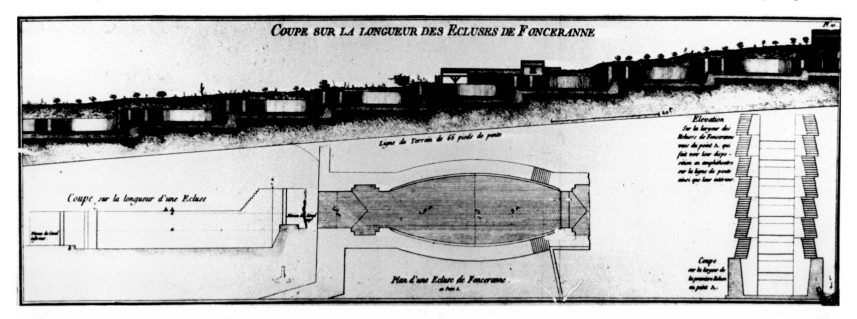

COUPE SUR LA LONGUEUR DES ÉCLUSES DE FONCERANNE

Coupe sur la longueur d'une Écluse

Plan d'une Écluse de Fonceranne

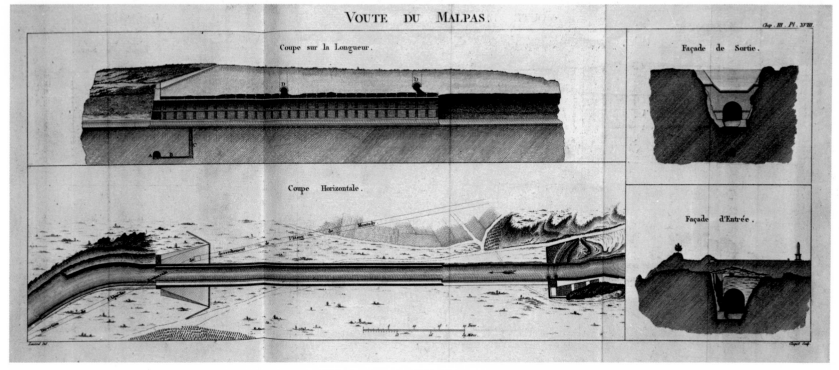

VOUTE DU MALPAS.

Coupe sur la Longueur.

Façade de Sortie.

Coupe Horizontale.

Façade d'Entrée.

Have, O Lord, compassion
on us and our town.
Already all depart one after another.
Your people wither away
as if under a spell.
There is no family
which has not lost a son or daughter.

There were also such terrible famines that the Estates of Languedoc were constrained to forbid the exportation of corn from the province during most of the 16th and the beginning of the 17th century.

In addition to these natural disasters, there were heavy taxes to pay, so heavy in fact that the people "pray to the King that, out of the goodness of his heart he should order the Parlement and the King's lieutenant in Languedoc not to put up prices nor to tax the corn, so that the country can retain its liberty, else grave dangers could ensue to the detriment of all the people."

The plains of Languedoc, more wooded than now, were the hideout of brigands. The forest of Bouconne, at the very gates of Toulouse, sheltered bands who came out to pillage, rape and torture. Decapitated bodies tied to trees were a common sight on the edge of the woods. Fear of brigands was part of everyday life and since the King's armies were warring in the Cerdagne and Roussillon, it came about that the soldiery quartered on a quiet village sometimes took it to be a conquered place. The peasants were not only forced to feed the army but were also subject to exactions of all sorts: "These warriors, instead of following the orders given by His Majesty concerning their food and upkeep, commit divers excesses on the local people. Not content with being fed at their expense, they exact large sums from the towns and villages through which they pass, pillage and steal, set fire to

the houses, rape the maidens and commit other grievous offences; all of which could cause great inconvenience to the service of His Majesty."

To add to their difficulties there were wild animals: "The suit presented by the Procureur du Roi concerning the lethal activities of wolves and savage beasts, which have led to the deaths of more than five hundred men, women and children in the seneschalship of Toulouse, has been noted."

As if all these problems common to most of their contemporaries in other parts of the Kingdom were not enough, the people of Languedoc brought down on their own heads new miseries. Always critical of central authority, whether religious or lay, and in spite of many Catholic missions sent to convert them, they eagerly embraced the Reformed or Protestant Church. "Together with Saintonge, Languedoc is the most heretical province of all." This new rebellion cost the province many more years of bloody repression, but one must not regard these religious struggles as pitched battles. "All these operations of war (the taking of châteaux, sieges of towns, rapidly undertaken and as quickly raised) are carried out by a very small number of bands led by determined and entirely unscrupulous Languedocian or Gascon peasants," wrote a contemporary.

Faced with the Huguenot problem, the attitude of the Estates of Languedoc was fairly equivocal; besides numerous ecclesiastics, there were Protestant barons and consuls. The populace, on the other hand, found itself hard done by, as is evident from this Catholic pamphlet called "The Gascon's prayer":

If Montpellier has good doctors,
we also have good surgeons
to draw blood from their hides.

The Trias aqueduct, which carries the canal over the river Cesse near Sallèles-d'Aude.
(18th-century drawing. Archives of the Canal du Midi.)

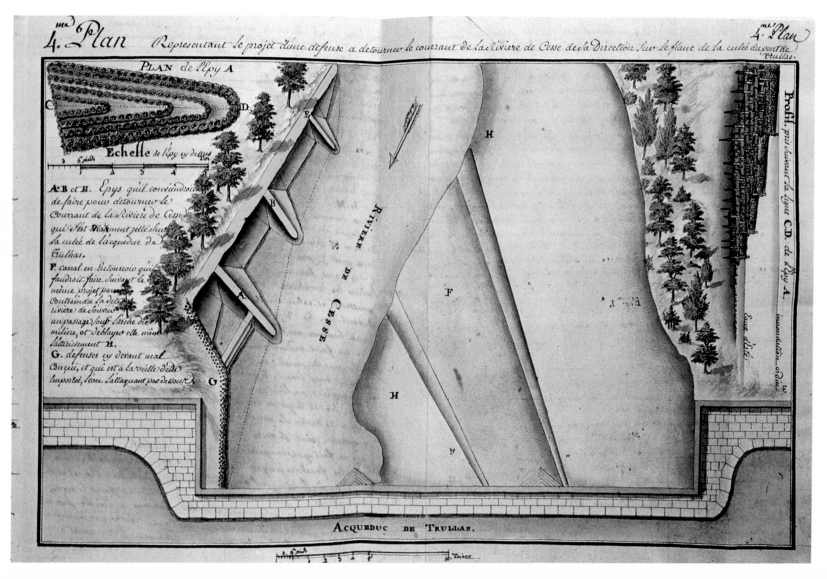

The Trias aqueduct. Where the terrain rendered
it necessary, bridges and aqueducts were
consolidated by foundations laid on wooden
piles anchored in the ground. (18th-century
drawing. Archives of the Canal du Midi.)

Plan for the modification of the course of the canal to allow barges to pass more easily under one of the bridges. (18th-century drawing. Archives of the Canal du Midi.)

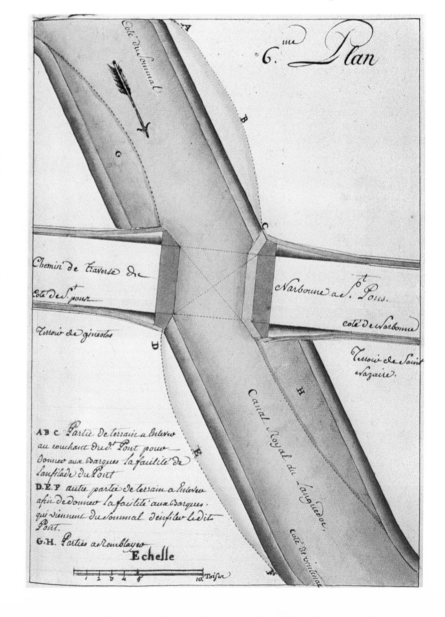

Response:
To the devil with all Huguenots,
exterminate them,
annihilate them.
Let proud Montpellier be humbled,
let Nîmes fall into the abyss
and Montauban blow away in the wind.

At all levels of society, the spirit of independence from the central power and also from the Catholic Church (the state religion) was a prevailing trait of the Languedocian character. The Protestants of the province did not even allow the Revocation of the Edict of Nantes to influence them. Admittedly the aristocracy did not set a very good example ; "Most of the nobles, even if they glorify themselves with the title of Catholic and vaunt their loyalty to the King, carry on the war half-heartedly and fraternize with the enemy."

The churchmen who occupied important posts were amply privileged, while the country curés were reduced to a state of appalling misery ; "That the simple curés are charged with the most laborious functions of the evangelical ministry ; that the care of the meanest members of the immense flock of Jesus Christ is in their hands, are indisputable facts. By what accident, or rather by what reversal of that order established by religion and demanded by reason and justice, are these most useful men treated with such little regard and condemned to all the horrors of an extreme and insupportable penury ?" It is hardly surprising therefore that the people avoided paying the tithes which contributed to the revenues of archbishops, bishops and abbots. A similar charge can be laid against the judiciary : "The clerks of the royal courts impose an infinite number of unjust levies on the people."

To be, at a time like this, Inspector General of the Salt Tax (*gabelle*) was certainly a lucrative post, but also a perilous one,

29

witness the misadventure which befell the Sire de Reignac in 1656. Despatched to Carcassonne by the Fiscal Court (Cour des Comptes) of Montpellier in order to reorganize and increase the *gabelle*, "he was besieged in his house and the populace sounded the tocsin all night, rolled a cannon in front of his door and threw handfuls of stones against his windows, with insolent cries of 'Long live the King, but down with the *gabelle*'". Pierre Paul Riquet had every opportunity to study the psychological situation of his province during the course of his many years as Intendant Général and King's prosecutor. During his many journeys through the land he had occasion to deplore the appalling degradation, bordering on slavery, with which the Mediterranean South was afflicted and its lack of communications with other parts of France, in particular the Garonne valley. Above Toulouse the Garonne was no longer navigable. Merchandise could only be transported on dreadfully difficult and dangerous roads, and commerce suffered accordingly. He could not forget his father Guillaume's condemnation of Arribat's project to construct a canal to link the two seas at the time of its rejection "without appeal" by the Estates of Languedoc. Many years had passed since then and Riquet had amassed a handsome fortune thanks to his lucrative post. He was the owner of splendid estates in Languedoc: Ramondens, in the heart of the Montagne Noire, and the château of Bonrepos, near Toulouse. Having weighed up the economic problems facing his province with all the maturity of his fifty years he now found himself facing the still unsolved geographical problems presented by the terrain. These had defeated the best technicians and Riquet was neither a geographer nor a specialist in the science of hydraulics.

The obvious line for a canal would link the rivers Garonne and Aude, via the watershed of the Stones of Naurouze. This watershed being situated at the foot of the Montagne Noire to the north, with the Pyrenees to the south, what was needed was to find enough water to fill the canal, which would only be of economic

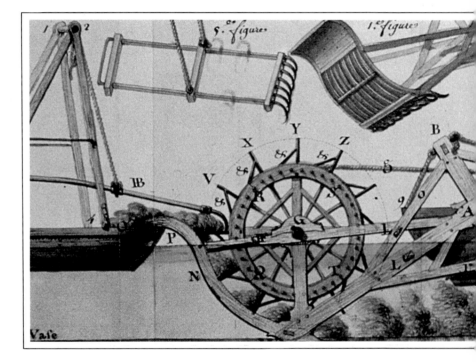

advantage to the community if in use all the year round. The Montagne Noire was the obvious source of water, but unfortunately it had an unreliable water table: during the summer months the streams had an insufficient flow for the project in hand. Riquet's stroke of genius lay in visualizing a channel (*rigole*) to collect the torrential waters of the Montagne Noire and divert them into the Sor. Another *rigole* on the plain would take the water thus collected to the watershed at Naurouze.

This had not, however, entirely resolved the problem of summer droughts, and Riquet — no longer working on his own, as Colbert had sent him scientific experts such as Messieurs de Bourgneuf and Vaurose and above all the Chevalier de Clerville,

Commissioner General of Fortifications — at first thought of creating fifteen or so small reservoirs along the trenches. De Clerville suggested instead a single large reservoir situated in the valley of Vaudreuille. This led to the construction, remarkable for those days, of the dam at Saint-Ferréol. A huge retaining wall of granite blocks, 871 m long, 35 m high and 5 m thick, dams the valley of the Audot, a tributary of the Aude. Above and below this embankment, the infill of rock and clay is held in place by two walls, the upper one completely submerged. At the base of these walls were constructed vaults for operational purposes and controlling the outflow of the water in the reservoir when the level falls below the sill of a sluice known as the "Badorque". A lower drainage tunnel empties into a channel cut into the bed of the Audot below the dam. An upper access tunnel leads to the stopcocks inserted into the masonry forming part of the dam wall and to be reached from the lower retaining wall. Above the dam itself and totally submerged are the two galleries nicknamed "the drum" and "hell". They are accessible only when the reservoir is emptied completely, which happens roughly every thirty years. The dam was put into operation in the year 1672 and remains in a perfect state of preservation to this day. It still arouses wonderment by virtue of the audacity of its design and construction and is considered a technical feat without parallel for the age in which it was built.

31

In the company of these expert technicians, Riquet, though not an expert himself, more than held his own. It was he who thought of employing a system of stopcocks of the kind used by wine-growers, instead of wooden sluices which would swell in the water. They work on the principle of two tubes, one revolving inside the other: "pipes of cast iron, as large as cannon".

Some years after Riquet's death, the King sent the famous military architect Vauban to Languedoc to inspect the canal. Vauban was full of admiration and declared that he would rather have achieved this glorious work than "anything he himself had done or might do in the future". He nevertheless carried out certain improvements, augmenting the waters of the Audot with those from the mountain *rigole* so as to prevent their being squandered in the Sor. Vauban extended the *rigole* from Conquet and diverted it to empty into the Audot above Saint-Ferréol, having crossed the Cammazes hills. A tunnel 123 m long was built to clear this obstacle, opening out into a deep conduit which discharges the water at the topmost point of the Audot's bank.

Once the problem of providing enough water for the canal was solved, Riquet and his colleagues had to turn their attention to the laying out of the canal itself, which was not without its own difficulties. Over a length of 257 km there are no fewer than 65 locks with a total of 103 sluices. Some of them are prodigies of engineering. The "ocean lock" which marks the beginning of the watershed leads to a section hewn from the solid rock. This is the place where the limpid waters of the high mountains are fed into the canal through a series of small channels. At Malpas the audacity of Riquet's thinking was realized in the construction of the famous tunnel on the approach to Béziers. Not to mention the splendid "stairway" of seven locks at Fonserannes and the majestic aqueduct spanning the river Orb, which leads to the very gates of

Béziers. It is almost as if Riquet was determined to provide a triumphal route leading to the city of his birth. Near the old town of Agde there is a round lock with a circular basin and three sets of lock gates, recently modernized. Depending on which gate is chosen, the barges can proceed either to the river Hérault, to the maritime canal which leads directly to the sea, or to the Canal du Midi itself.

But let us now leave descriptions of individual architectural works along the canal. Everything is harmonious there, from the humblest oval lock and the charming lock-keepers' houses in pure 17th-century style to the breathtaking view on arrival at the lagoon at Thau with the ever-changing colours of its waters.

Riquet's ambitions were not perhaps quite realized to the full, since the width and depth of the canal meant that the "linking of the two seas" only applied to fairly narrow barges, but he nevertheless imparted to it a remarkable lushness. From Ponts-Jumeaux at Toulouse to the lagoon at Thau, 45,000 trees were planted to line the banks of the canal. The different types of tree chosen testify to a wide knowledge of the climates of the different regions through which the canal passes. Poplars, willows and elms were chosen for Aquitaine, pines and mulberries for Languedoc, cypress and olive for the Mediterranean. The olives and mulberries have now disappeared and the poplars have been replaced from time to time by plane trees, whose leaves — which are slow to rot — choke the bottom of the canal in autumn; the pines remain. Each type thus marks the invisible boundaries of the climatic regions and filters the light in its own particular way. This is not the least of the charms of the canal.

Riquet, the great adventurer, has left us a wonderful heritage to be valued for its technical perfection and for the way in which it is integrated into the countryside. The Midi is proud of its canal!

The threatened canal

ALTHOUGH THE GARONNE has always been considered navigable from Roquefort-sur-Garonne to its estuary, the irregularity of its flow prevented the passage of heavily laden vessels. Vauban was already considering this problem at the time of the inauguration of the Canal du Midi. He thought it possible to construct a canal parallel to the river, thus forming a navigable route from Sète to the Gironde.

It was almost two centuries later that this project came to fruition and the Canal Latéral, begun in 1838, was finally opened in 1856. It is 193 km long with 53 locks, and the banks are planted with poplars. For much of its length the countryside around is beautiful and the sight of the river rolling majestically along past stony-beached islands, flanked by its canal and by hillsides reminiscent of Tuscany, is truly a splendid one.

For long periods the canal was "blacked" by the boatmen. No sooner was there sufficient water in the Garonne than the barges abandoned the canal and took to the river. It took less time to complete the journey in this way and there were no tolls to pay; also, since the 16th century many of the Garonne's tributaries (Lot, Drot, Tarn, Baïse, Agout) were used for navigation, as vestiges of locks and weirs testify. All the same, this canalization of the Garonne greatly increased commercial traffic in the long run. From now on all large-scale consignments of merchandise would go by water.

Towards the end of the 19th century there was intense activity on the Bordeaux-Sète route. Just as in our present age the lorries thunder down the motorways, so at that time there was an unending stream of boats carrying building materials, and oil, cereals and wine from the Gironde, the Tarn or Bergerac, back and forth along the canal. As a result, life along its banks increased by leaps and bounds. In the towns commercial ports sprang up, their quays humming with activity. Among the barrels and carts a swarming crowd of dockers, stable-boys, sailors, carpen-

278. - AGEN. - Sur le Pont-Canal

The roadsteads of Bordeaux. At the beginning
of this century more than 250 open and some
decked boats regularly made the journey
between Bordeaux and Sète. The boats with
square prows moored side by side like rafts
served for the storage of dried cod.

H. L.   BORDEAUX - La Rade

Sète (Cette), at the other end of the canal. Life around the lagoon at Thau was extremely busy. The tugboats of Sète (lower photograph) were essential for the crossing of the lagoon. Their owners formed a corporation well known for its exorbitant tariffs.

ters and carters kept up a constant barrage of noise which vied with the whinnying of horses and the shouted orders of the foremen.

As well as ordinary commercial traffic there were the mail boats. Run by the canal administration, they could do the journey from Toulouse to Sète in thirty-six hours, with regular stops along the way so that passengers could connect with the railways or with the steamboat to Bordeaux, which used the Canal Latéral, or the one which crossed the lagoon at Thau. In 1834 the Duc de Caraman introduced the "express boats" which made the journey between Toulouse and Beaucaire in either direction in 115 hours. They were decked boats which carried 60 tonnes of cargo as well as passengers and were cheaper than the mail boats.

Unfortunately, after the nationalization of the canal, both these types of boat disappeared, leaving only those of the private contracting firms, which is all that remains today.

Adapting naval architecture to the needs of canal users resulted in barges with long upswept frames and truncated sterns to accommodate the rudder and house the living quarters. Of Mediterranean design, these boats belong to the same family as the "allèges" of Arles and the "Bateaux boeufs" of Sète, so called because they were drawn by a pair of oxen.

The name Beaucaire crops up again and again when recalling the years of the canal's greatest activity. Long before the opening of the Canal du Midi, there was already a waterway between the lagoon at Thau and the Rhône. By this route salt from the salt-pans at Peccaïs near Aigues Mortes reached the markets of France. The town of Beaucaire, strategically situated opposite Tarascon (the military key to the Rhône valley) was in its way as important to that river as Bordeaux was to the Gironde. Ships from all over the Mediterranean assembled here, and it was the place where they met the small river craft of the Rhône; the great

160 – CETTE - Station Balnéaire - Pêcheurs à la Ligne

4006 – Cette   La Pointe courte - Entrée de l'Etang de Thon

35

Typical canal barge designs from the 18th century to today. The last of the wooden boats still operated until a few years ago. They have now all been replaced by steel-hulled engine-driven craft.

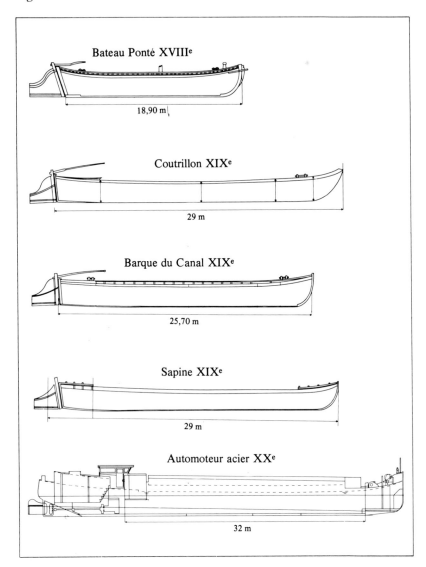

Bateau Ponté XVIIIe

18,90 m

Coutrillon XIXe

29 m

Barque du Canal XIXe

25,70 m

Sapine XIXe

29 m

Automoteur acier XXe

32 m

fair, held at Beaucaire in July, saw the mingling of merchants and seamen from Spain, Italy, Greece and Turkey.

What better place could have been chosen for the grand opening of the canal in May 1681, than this great commercial arena, at the conclusion of a long and dignified voyage?

Three centuries have now gone by since the opening of the canal. To travel by inland waterway today is to leave behind the stress and strife of modern life, but we should not allow ourselves to be lured into a mistaken nostalgia. The slow speed imposed by this type of water travel seems to petrify the rhythm of those who work on the boats or on the canal banks. One would need a magic ring to awaken the canal from its enchanted sleep which is its charm, but also its nemesis.

Despite the difficulties posed by this way of life, there are many young men who would like to follow in their parents' footsteps. Most of the present-day boatmen are the sons, grandsons or even great-grandsons of boatmen. The Larrose family, for example, take pride in tracing their descent from a bargee of Riquet's time. They have a very clear idea of the unconventionality of their calling. In their capacity as masters of their own vessel they feel free of all constraints. They work hard at their job, struggling to make ends meet and to preserve the viability of their chosen occupation; but the problems have multiplied over the past few years. The purchase of a barge represents an enormous capital outlay for the owner-operator. The price of fuel, the speed and economy of rival road and rail transport, as well as the slow decay of the canal itself, have obliged all but the most determined to give up. Cargoes other than wine or cereal crops are rare in spite of the low cost element. In an age devoted to speed it seems paradoxical to waste five days between Bordeaux and Sète. What is more, the small size of the locks, which for the most part has not been increased since the 17th century, and the shallowness of

Port Saint-Sauveur, Toulouse, about 1918. Barge Nº 242, a grain transporter, being unloaded. At this date the canal boats were still identified only by number. Names did not come into general use until the 1920s.

Toulouse. A steam barge near Matabiau railway station. The Garonne and the Pont Neuf at Toulouse.

the canal, which does not allow for the deeper draught of the long boats already working on the other canal systems of France, is creating an impossible situation which will kill present-day commercial traffic if not remedied in the very near future.

The boatmen are aware of all this; it is after all one of their daily preoccupations. But for the most part they like their way of life, which is a curious combination of sedentary and nomadic elements. They live on board with wife and children, at least when the latter are small, and they usually have relatives in one of the towns or villages along the canal — places such as Castets-en-Dorthe or Agen, Toulouse, Carcassonne or Sallèles-d'Aude.

The living quarters of the boatman and his family are in the forward part of the barge, the aft housing the wheelhouse and controls, the galley/dining room and a second bedroom with ablutions close by. The interior is decorated with taste and style, just as it would be on land. Family festivals and the rest period between journeys are spent at the town where the other members of the family live.

The boatman rises at 6 or 7 a.m., depending on the time of year. From time to time the need to be first in line for the next lock obliges him to get up even earlier, as a voyage with cargo takes longer than when the boat is in ballast. After a hasty breakfast the daily shift begins. The tiller can never be left unattended, for the route is both winding and narrow.

On arrival at a lock — the gates have been opened in advance, commercial traffic having priority over all other canal users — the boat's speed is reduced. Nosing the bow gently into the lock, the boatman throws a rope to the lock-keeper, who winds it a couple of times round a bollard to halt the barge's forward movement. The lock gates close behind it and the water is allowed to flow in, the lock-keeper being assisted by the boatman in manoeuvring the heavy sluice gates. This is the moment for an exchange of news: "Your brother-in-law is coming down tomorrow with the 'Tage'"; "The 'Ben Hur' is laid up at Toulouse with rudder problems"; "Did you know that Jean-Luc is giving up at the end of the year?"

From lunch-time until 7 p.m. in summer (6 p.m. in winter), the journey continues, the radio playing softly, the boatman, his eye constantly on the bow, correcting the course from time to time. When the canal closes down for the night, the barge ties up in readiness to enter the next lock first thing in the morning.

The boatmen are united by a deep feeling of corporate solidarity; all have the same problems and preoccupations. Engineers and "technocrats" who speak in ringing phrases about the future of the canal are often criticized for not really understanding how it works. Not having to hurry, the bargee is able to relish every passing hour and the changing countryside, which he knows like the back of his hand. He watches out for the places where navigation is difficult and slumbers when cruising along the straight stretches of canal, sitting on the helmsman's stool with his feet resting on the tiller. No two days are alike; each proves a little different from the last. What peace! What freedom!

Another breed of happy men are the lock-keepers, often related to bargees or themselves retired boatmen. Haven't we all envied them their snug little houses in pure 17th century style? The work is demanding, however, particularly in summer, when the pleasure-boat traffic is at its height. It needs a fit man to do this job, although those passing through give a helping hand. In winter the lock-keepers have more time on their hands, which they occupy with fishing or collecting old souvenirs and postcards of the canal in times gone by. Nor are the lock-keepers' wives idle, helping as they do to operate the sluices and participating in the to-and-fro of canal life, as well as fattening up the geese and

39

cultivating the kitchen garden whose produce is sold to passing boats. The unceasing summer traffic of the pleasure boats is amusing and instructive to her. She is happy to invite you into her home to taste a *pastis* or a "carthagène" and is equally happy to try your Flemish beer and learn the odd word of English.

The people who work on the canal form part of one large family, with its own code of ethics, its own ties and its own language. The common denominator is found in a mutual interest in the canal and its ways.

There are also those who bicycle along the towpath or wander on foot in the shade of the plane trees; the fishermen who dream away the time without caring very much whether or not they catch anything. The banks of the canal in Toulouse have been so well landscaped that it is a pleasure in summer to stroll along to the locks and watch the boats pass.

They say that within the breast of most men resides a sailor, and the popularity of cruising on the waterways between the two seas bears this out. There are several companies which hire out comfortable barges and cruisers. It is not necessary to be an accomplished sailor. Everything has been organized so that the "skipper" and his crew/family can get away from it all for a week or two and discover "nostre païs" and "nostre canal" at the restful speed of 6 km per hour. Nor should we forget the true sailors from the North : English, Dutch, Belgians and French-men from the Atlantic ports. For them a canal trip often serves as the ideal prelude to an ocean cruise.

The dilemma : historic monument or modern transport system ?

PIERRE PAUL RIQUET'S 300-year-old Canal du Midi must not be allowed to become simply an historic monument. Its modernization is indispensable to the commercial health of southwest France.

There is much talk these days about the development of the Midi-Pyrénées region, but it is essential that inland waterways play a part in this process. It is a modern fallacy to call by the name "Canal des Deux Mers" what is in fact three separate canal systems covering between them the 530 km that separate the Atlantic from the Mediterranean. They are the Canal Latéral parallel to the Garonne, between Bordeaux and Toulouse, built between 1838 and 1856; the Canal du Midi proper, built, as we have seen, between 1667 and 1681, and completed by the junction at Port-La-Nouvelle; and the canal from Sète to the Rhône, completed in 1820, which was destined to be of enormous importance in opening up the waterways of the South to the trans-European Rhine-Rhône network.

The region in question is in fact closely linked to this waterway since the Canal Latéral runs through it in the départements of Tarn-et-Garonne and Haute-Garonne as far as Toulouse. There the Canal du Midi itself begins and crosses the Toulousain countryside until it reaches the Mediterranean. Any development of the region by means of the canal network should not, however, be limited to meet local needs only. It can only work if a regular traffic can be started up between the southwest and the whole of the rest of Europe.

The First World War spelt the end of the prosperity and heavy traffic of the 19th century. After a short revival when hostilities had ceased, business flagged again in the 1920s. After the end of the Second World War, the freight carried amounted to a maximum of about 500,000 tonnes per annum, cargoes consisting mainly of cereals, wine, foodstuffs, wood and coal, scrap-iron, building materials and fertilizers.

During the 1970s the level of traffic was further reduced, partly owing to the increase in fuel prices, though this is not the main reason. The Canal du Midi is way behind in modernization compared to the rest of the French canal system. From 1970 to 1973, forty-eight of the fifty-three locks of the Canal Latéral were increased in length to 40 m and, at Montech, a series of five locks was replaced by a "pente d'eau" (water slope), a new concept in canal engineering. On the other hand the Canal du Midi is no more than 1,80 m deep in certain sections and its locks are only 30 m long, thus restricting its use to barges carrying 150 tonnes. It is therefore not possible to convey cargoes directly from Bordeaux to the Rhône, as the passage from one canal to another would necessitate a change of boat and the transference of the cargo. This is uneconomic in view of the time involved and the high cost of manual labour.

All these factors are reflected in the following unhappy statistics: in 1975 the coopérative de la batellerie (Inland Waters Cooperative) owned 130 barges; by June 1978 it had only 60. Barges are sold for conversion into floating restaurants or night-clubs, or they simply disappear and their boatmen with them. Sometimes one sees them moored; they have become the floating homes of boatmen who were forced to leave the calling. At least these men have the consolation of living in a port and watching the canal traffic go by. There is only one remedy: the complete modernization of the Canal du Midi to allow the passage of barges carrying up to 350 tonnes. This modernization should include the lengthening of all locks to a standard 38.50 m and an increase in water depth to 2.50 m.

As pointed out, modernization of the Canal Latéral began in 1970 and boats with a length of 38.50 m can now use it, carrying loads of 250 tonnes. Once the "pente d'eau" at Mon-

41

Castelnaudary. Skating on the "Grand Bassin"
during a very hard winter, January 1914.
The port at Castelnaudary.

Castelnaudary (Aude) — Janvier 1914 — Patinage sur le grand Bassin

tech has been deepened to 2.60 m, it will be possible to convey loads of 350 tonnes along it. If similar work had been carried out on the Canal du Midi at the same time, this would have meant an increase in traffic of around 20 per cent.

In June 1977 a three-year modernization plan was put in hand at a cost of 70 million francs, subsidized 60 per cent by the State and 40 per cent by the regional councils of Aquitaine, Midi-Pyrénées and Languedoc-Roussillon. By the end of 1979 this plan had been carried out to the letter.

By June 1980 two new sections of the old canal were able to take 38.50 m barges: in the west the Toulouse-Baziège section, and in the east, the section from Sète to Béziers and the La Nouvelle junction. In addition, the plan to replace the seven locks at Fonserannes by a single new lock at a cost of 27 million francs is already in progress. Needless to say, the old locks will be preserved as a magnificent example of the engineering works of their time. In fact conservation of the 17th century engineering works of the old canal is an integral part of the modernization programme and includes plans to "double", rather than destroy, old bridges and bypass old locks, even though this will add considerably to the cost of the work. The future of Languedoc depends on the carrying through of these modernization schemes.

When Frenchmen in the South discuss the future of their region, they cannot refrain from speaking with some bitterness about the way the boatmen's calling has been allowed to decline, with the result that much heavy industry which would otherwise have been attracted to this area has gone elsewhere, on account of the high cost of road and rail transport. A modernized canal might have forestalled all this.

In November 1978 *La Dépêche du Midi* published an appeal on behalf of the boatmen and their jobs, after learning of the

Castelnaudary. The bridge and windmills at Saint-Roch, 1902. Until the 1920s, windmills were part of the canal scenery and reflected the economic life of the area. Originally used for crushing pastel or woad for dyeing, a product which made the fortune of the Lauragais in the 15th and 16th centuries, they were later converted into mills for grinding corn.

Potter at work in Castelnaudary.

problems of one of their number whose barge had run aground, blocking the canal, and who had to pay heavy fines. "In a world where the poetry of life is fading like a dying flower", said this newspaper, "the Canal du Midi is for some their last retreat. It would be too easy to obliterate with a stroke of the pen the economic role played in our region by this waterway, under the pretext that it is an outmoded form of transport. The Canal du Midi, to some extent by the very fact of the modernization work now in progress, raises very great difficulties for those connected with it, from Bordeaux to Sète. If the necessary decisions are not taken urgently it will mean the end of the boatmen's calling and livelihood. Let us therefore say, No! The boatmen of the Midi must not be condemned to shipwreck. . ."

At the same time, the success of the pleasure-boat industry, often run by foreign companies, is all too evident and resented by some of the locals. Is this a reawakening of the old martial spirit of the southerners threatened by invasion from the North? No; France is one country now, forged by centuries of trials suffered in common by North and South. The plan for the "Aménagement du Grand Sud-Ouest", already under way, has given hope to those who live in this most attractive area. Their hope for the future is that the voice of the Southwest will be heard in the corridors of power just as loudly as those of other regions geographically nearer the seat of government.

Since 1898 the Canal du Midi has belonged to the State. Let us set down briefly its salient features and the events leading up to France's historic legacy. The canal is 240 km long by 16 m wide at the surface and 10 m on the bottom, 2 m down. It is connected with the river Garonne by the Canal de Brienne, upstream from Toulouse. The watershed is at Naurouze, 132 m above sea level. It has 64 locks: 39 single, 19 double, four triple, one quadruple (at Castelnaudary) and a succession of seven (at Fonserannes). The cost of the work in its entirety was 15,249,399 livres, of

44

which 7,484,051 was contributed by the King, 5,807,831 by the province of Languedoc and 1,957,517 by Riquet from his own pocket. It took the latter's descendants more than a century to pay off the debts thus incurred and in 1784 they regained possession of the canal. At the Revolution they emigrated and their right to it was rescinded.

In 1810, under the First Empire, three-quarters of the stock of the company was distributed by Napoleon to retired soldiers with a record of meritorious service to their country. At the restoration of the monarchy, the Riquet family retrieved part-ownership of the canal. In 1858 the "Compagnie des Chemins de Fer du Midi et du Canal à la Garonne" took over, and in 1898 the State bought the canal, which thereafter became national property.

The three-hundreth anniversary of the Canal du Midi coincided, by chance, with "L'année du Patrimoine". Has this perhaps some special significance? Louis XIV's century bequeathed an enormous wealth of monuments and works of art. Versailles attracts thousands of visitors every year. The architect Vauban left us his remarkable fortifications throughout the country. Curiously enough, the Canal du Midi, probably the largest engineering undertaking of its time, but built strictly for economic reasons, has been overshadowed by more spectacular monuments. The ever-increasing numbers of pleasure-boats that take advantage of its facilities each successive year is a measure of its "rediscovery".

The celebration of the tercentenary has contributed to the renaissance of the canal. Let us travel along this waterway, seeing with new eyes every lock, every bridge, every lock-keeper's house, every village, and bearing in mind that three centuries have passed, during which the stones of the building have acquired a warm patina and the noble trees the dignity of age, and that this is a very precious heritage. Let all those who,

Castelnaudary. Mid-Lent fair, 1924.
Horse-drawn barge loaded with barrels in the lock at Vivier.

45

Sète. Sailing-boats with barrels of wine ready to
be put aboard.

Edition A. F.

Cette — Voiliers séchant leurs voiles

by reason of their job or simply in the course of a pleasure cruise, have been able to taste the unforgettable charm of this journey back into time, realize that if it could, the canal itself would cry out, not only for its survival but also for its rejuvenation.

The need to bring about such a rejuvenation on economic grounds has been expressed countless times, but there are subtler reasons to do with France's artistic and architectural patrimony as represented by the ports of the Canal du Midi, such as Moissac, Toulouse, Castelnaudary, Carcassonne, Béziers and Adge — to name but a few of the most important — and these should encourage the powers-that-be to make the dream of Pierre Paul Riquet and of the whole of the Midi come true at last.

Odile de Roquette-Buisson

From sea to sea

The roadsteads and port of Bordeaux. *(Preceding pages and left.)* Like most estuary cities Bordeaux is both a maritime port and the great terminal port of the river Garonne, meeting-point of all the boats of the Midi, from Provence and from the Rhône, which traverse the 560 km from Bordeaux to Beaucaire. Of the sizeable fleet which sailed between the two seas at the beginning of the century, there are now only about 60 boats using the Canal Latéral even though this has been modernized, and scarcely ten that use the Canal du Midi, whose antiquated state does not allow the passage of barges more than 32 m long.

The Garonne above
Castets-en-Dorthe.

Castets lock, the
starting-point of the
Canal Latéral. It is at
Castets-en-Dorthe that
boats coming from
Bordeaux encounter the
first of the locks.
Modernization began in
1970 and is now
finished on all 193 km
of the canal. The new
locks are all 40 m long
by 5.50 m wide and
2.20 m deep, allowing
the passage of boats
with a length of
38.50 m as far as
Toulouse.

The Garonne below Castets-en-Dorthe.

Lock-keeper's house at Castets. The graduated scale on the wall shows the levels of the river's highest floods. Along its lower reaches, from Castets to Bordeaux, the Garonne has always been navigable. Sailing barges and lighters from the Gironde, as well as the canal boats, were towed along the river by steam tugs. This 53 km section is always busy with canal barges and pleasure-boats. It is not the most beautiful part of the journey but nevertheless offers travellers a pleasant prospect with a number of interesting places such as Langoiran, the château of Cadillac and the vineyards of Saint-Macaire and La Réole.

Castets-en-Dorthe: the canal's first lock and an
important port for shipping. Many boatmen's
families live here and it is a popular place to
stop and spend time relaxing or between jobs.

Moissac. A footbridge in early 19th-century style. The Canal Latéral passes through Moissac at the 63 km mark, after spanning the Garonne via the spectacular Agen aqueduct. Everyone interested in Romanesque architecture makes a point of visiting the church of Saint-Pierre, whose porch and cloister are among the finest in France.

(*Overleaf*) The Ponts-Jumeaux at Toulouse. At this point the Canal Latéral comes to an end and the Canal du Midi begins. Here a stretch of water extending over 350 m, three bridges spanning three separate waterways, and François Lucas's monumental relief together make up a marvellous piece of canal architecture. Alas, it is now disfigured by a recently constructed network of urban motorways.

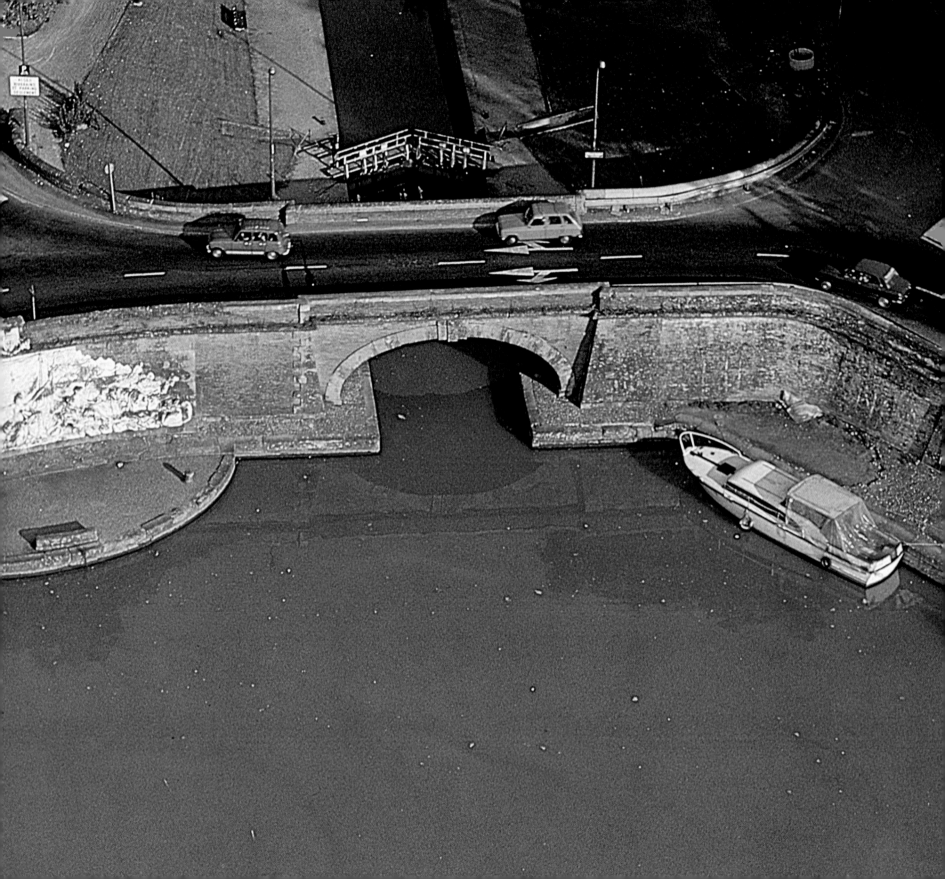

Toulouse. Port Saint-Sauveur. Until the early years of this century, this port was one of the largest for small craft in the Midi. Barges competed for berths and a whole neighbourhood of the town was influenced by canal life. Pleasure-boats still fill the port, but barges and boatmen can now be counted on the fingers of one hand.

Toulouse. The canal in the centre of the city, near the railway station (Gare Matabiau) and the port of Saint-Etienne.

The route of the canal through Toulouse was set at the northern limits of the 17th-century town. Today it is surrounded by modern

Toulouse and has given the city a magnificent waterside boulevard, which town planners have not always respected.

Boatmen and their families, fishermen and pleasure-boat owners in Toulouse. Most of the Toulouse boatmen have now disappeared, although some who have given up their calling still live on their boats, now fitted out as houseboats. Fishermen on the banks of the canal are very much part of the local scene, while strolling along the towpath is an increasingly popular pastime.

Toulouse. A lock on the Canal de Brienne and Port-Sud, in winter. At the southeast exit from Toulouse, Port-Sud has been reorganized since 1973 as a marina for pleasure-boats ; many of them are tied up here for the winter, awaiting the hot summer weather to slip down to the Mediterranean.

The lock-keeper's house at Emborrel. Most of the Canal du Midi's lock-keepers' houses were built in either the 17th or the 18th century, to an identical design. They have been the homes of veritable dynasties of lock-keepers.

A laden barge enters the lock at Saint-Roch.

Lock-keeper's house, built in 1752, at the "écluse du Sanglier" (Lock of the Wild Boar) near Baziège.

A cyclist climbing the towpath stairs at Saint-Roch lock.

Montagne Noire, the lake of Lampy and the waterfalls at Saint-Ferréol. Without the Montagne Noire the Canal du Midi could never have been built. Riquet's stroke of genius lay in the decision to channel the waters of the mountain to Naurouze, the watershed between the two seas. The reservoir at Saint-Ferréol, and later those of Lampy and Cammazes, were installed to regulate the supply of water for the whole canal system. The dam at Lampy was built in 1777 to serve as a reservoir for the branch of the canal from Sallèles-d'Aude to the river Robine at Narbonne. The waterfalls at Saint-Ferréol reunite the waters of the upper Laudot with those of the mountain *rigole* or trench.

Montagne Noire seen from the slopes of the Lauragais.

The dam at Saint-Ferréol, constructed between 1667 and 1672, is 800 m long, 35 m high and 120 m thick at the base. Built of granite ashlar, it spans the narrow valley of the Laudot and is the best known of all the works along the entire canal system devised by Riquet.

The banks of the trench on the plain and those of the canal itself are the favourite haunts of sportsmen. A pleasant stroll in the autumn sunlight makes up for the scarcity of game.

The "Poste du Laudot" at the foot of Montagne Noire marks the place where the waters from the trench (or *rigole*) on the plain mingle with those from the middle Laudot. This station plays a very important role. If there is too much water, an overflow diverts the surplus to the lower Laudot; if there is too little, more is provided from Saint-Ferréol.

(Overleaf) The "Seuil de Naurouze" in winter. In the foreground is the obelisk marking the highest point of the watershed between the Mediterranean and the Atlantic, 190 m above sea-level. Behind is the octagonal stone-built reservoir, later abandoned because of subsidence.

Guinea geese. Unlike the grey goose, they cannot be fattened to produce *foie gras,* but are often kept by lock-keepers.

In the lock chamber, as the water flows out through the lower sluices, the stone sides become impressively high damp cliffs, shutting out the daylight.

Castelnaudary;
18th-century bridge at
the entrance to the dock
basin.

The port of
Castelnaudary. These
days pleasure-boats
have replaced
commercial traffic and
give renewed life to the
otherwise deserted
quays.

Castelnaudary; view of the town across the great basin. The Chauriens (inhabitants of Calstelnaudary) were aware of the benefits to be gained from running the canal through their town. It was at their request and expense that Riquet diverted the canal and built the dock basin. The capital of "cassoulet" owes a new-found prosperity to the proximity of the canal. The great basin is now the home port of the "Blue Line", one of the companies which organize cruises on the canal. Their familiar blue-painted houseboats are now part of the everyday scenery of the canal.

The "Nemo", an old working barge converted
into a pleasure-boat by an English couple,
docking at Castelnaudary.

Castelnaudary market. All along the canal, as
elsewhere in the South, each town has its
market day.

The quadruple lock at Saint-Roch at the exit from Castelnaudary.

Plane trees line the canal on the Saint-Sernin reach. Between Toulouse and Carcassonne the oaks, elms and limes planted when the canal was built in order to provide shade and reduce the drying out of the banks, have slowly given place to plane trees whose almost indestructible leaves pile up on the canal bed and block the lock gates.

95

Pleasure-boats in Vivier lock. To cruise along the canal in your own boat or a hired motor launch is becoming more and more popular. On some summer days there is an unending stream of boats waiting at the locks. Smaller boats are allowed in in twos or threes, strictly on the principle of first come, first served, except for commercial barges which have priority over all other craft. A canal cruise is not as restful as the calm waters of the canal might suggest. The actual navigation, which includes observing the rules when passing other boats and manœuvring one's own through locks and under bridges, stocking up with food, as well as the general maintenance of the boat, make it a busy time for all on board.

(*Overleaf*) Three sailboats approaching Guillermin lock.

Before passing through the vine-growing départements of Aude and Hérault, the canal traverses the Lauragais, the territory that links Aquitaine with the Mediterranean region. This has been the traditional granary of the South for many centuries.

The canal wending its way between Villesèque and Carcassonne. Contrary to the usual conception of a canal, the Canal du Midi rarely runs in a straight line. In order to limit the cost of terracing, Riquet followed the natural contours of the ground as often as possible.

The river Aude and Carcassonne. Unlike the Chauriens, the Carcassonnais were not prepared to pay the cost of diverting the canal, which therefore bypasses the town to the north.

The provision of public wash-places was one of the many sources of income for the canal administration. Nowadays there are scarcely any washerwomen and therefore these wash-places are virtually unused.

Trèbes lock and its mill. Wherever there was a sufficient gradient, Riquet installed a mill in order to make best use of the hydraulic power provided by the canal. The revenue thus produced was another source of income contributing to the upkeep of the canal.

In hot weather, the canal banks offer a peaceful and shady spot for a picnic. The meal over, the men try their hand at a little fishing while the womenfolk gossip and keep an eye on the children.

When two boats belonging to the same family meet, the engines are cut and the latest news is exchanged.

(*Overleaf*) The château and village of Argens, seen from the canal.

The "Lucre" arriving in ballast at Sallèles-d'Aude. The handful of barges still operating on the canal owe their survival to the wines of the départements of Aude and Hérault, which are taken on at Sallèles and transported to Bordeaux.

A lock-keeper turning the handle that opens the sluices. In the summer months the lock-keeper's work is increased tenfold by heavy holiday traffic. There is, however, an unwritten law that those using the canal lend a hand in operating the locks.

The Pont des Marchands at Narbonne. This bridge spanning the Robine is one of the few old bridges in France to have retained the houses built on it.

The Robine is a tributary of the Aude, linking Narbonne and Port-La-Nouvelle. Canalized as early as Roman times, it was re-adapted in the 18th century to give a second outlet to the sea.

The lagoon at Bages and Port-La-Nouvelle.
Between Narbonne and Port-La-Nouvelle, the
Robine canal crosses the narrow spit of land
separating the Bages and Ayrolle lagoons.

"Pont Riquet" near Homps.

Scene near Capestang. Here the canal approaches the département of Hérault, the world's largest vineyard, whence come the Frenchman's traditional "gros-rouge" and, latterly, the light and fruity "vins de pays".

Few of these
17th-century bridges
with semicircular
arches, called "Ponts
Riquet", still exist. They
are too narrow to allow
the new 5 m-wide
barges to pass and are
all scheduled to be
enlarged when the final
phase of the canal's
modernization scheme
is put into effect.

The wheelhouse of the
"Ben-Hur": one of the
last of the Canal du
Midi's wine
transporters.

Beyond Capestang the canal skirts the Ensérune hill, a rocky outcrop in a sea of vines stretching as far as the eye can see. Now covered with cypresses, pines and olives, it is a place where archaeologists have discovered traces of pre-Roman civilizations.

The dried-up lagoon at Montady, to the north-east of the Ensérune hill, is a little-known curiosity of the Mediterranean south. The perfect radial design of the fields surrounding the central circle resulted from the construction of drainage ditches to empty the lagoon.

The sequence of seven locks at Fonserannes is undoubtedly the most spectacular engineering feat on the canal. This aquatic "stairway" which follows a 54 km stretch without a lock, has, in the form of the Cesse trench, its own water supply. There is a plan to bypass the old locks by means of a single huge lock, which is opposed by environmentalists and ecologists.

The aqueduct over the river Orb, a few hundred metres from Fonserannes, makes the approach to Béziers appear almost triumphal. It was built during the reign of Napoleon III.

The old town of Béziers, beside the river Orb, apart from being the wine capital of the South, prides itself on having been the birthplace of Riquet. It is also known for the remarkable fortified cathedral of Saint-Nazaire, built in the 13th and 14th centuries, and the seven-arched bridge, one of the most important medieval bridges in France.

At Agde, Riquet built the well-known round lock which allows boats to make for either the river Hérault or the sea. It has recently been modernized and can now take the new larger barges.

A sailing boat by the locks at Fonserannes.

(Overleaf) Agde: the fishing port on the Hérault.

Sète: the "joutes sétoises" or water tournament. The terminal point of the Canal du Midi and a busy fishing and commercial port, Sète is entirely Riquet's creation. He constructed the port and it was the activity on the canal that brought prosperity to the town. Its charm lies in the streets beside the water where cars and boats move in close proximity. Summer is the time for the tournaments. Local champions converge on the town from the surrounding countryside to try their skill with lance and shield; and Sète takes on the appearance of a little Venice.

The Thau lagoon; pink flamingoes and the light-house at Les Onglous, where the Canal du Midi ends. Unlike the other large but shallow lagoons of lower Languedoc, this one is navigable over its entirety. On the far side is the Canal de Sète leading to the Rhône.

For their help in creating this book, thanks to:

Inès and Diego

Robert Barbereau, Archivist of the Canal
André et Simone Bos, Collectors
Jacques Cabanes, Pilot
Vincent Calduch, Reproduction and documents
Pr. Roger Camboulives, Documentation
Louis Cazaneuve, Collector
Catherine Coustols, Archivist of the Ecole des Beaux Arts of Toulouse
Odile Fournier, Proprietor of the château de Bonrepos
Fernand Gradit, Secretary general of the *Consortium des voies navigables du Midi*
Jeanne Guillevic, Conservateur du Musée Paul Dupuy in Toulouse
Gabriel Houlie, former Ingénieur en chef of the Canal du Midi
M. et Mme Huguet, Collectors, boatmen of "La Reole"
Jean-Luc Lacampagne, boatman of "Ben Hur"
Alain Lefebvre, Architect
Jacqueline d'Orgeix, Documentation
Isabelle Roquebert, Documentation
Henri Sarramon, President of the *Consortium des voies navigables du Midi*
Sambona Tep, Musée Paul Dupuy in Toulouse

Air-Inter company

The photographs on pages 72-73 and 90-91 are by Isabel Lefebvre
The book was designed by René Schumacher